United States
Environmental Protection
Agency

UPDATED 2016

Release Detection For Underground Storage Tanks And Piping: Straight Talk On Tanks

EPA 510-K-16-003
May 2016

EPA wrote this booklet for owners and operators of underground storage tanks (USTs).

This booklet describes the 2015 revised *federal* UST regulation. Many states and territories (referred to as states in this booklet) have state program approval from EPA. To find a list of states with state program approval, see www.epa.gov/ust/state-underground-storage-tank-ust-programs.

If your UST systems are located in a state *with* state program approval, your requirements may be different from those identified in this booklet. To find information about your state's UST regulation, contact your implementing agency or visit its website. You can find links to state UST websites at www.epa.gov/ust/underground-storage-tank-ust-contacts#states.

If your UST systems are located in a state *without* state program approval, both the requirements in this booklet and the state requirements apply to you.

If your UST systems are located in Indian country, the requirements in this booklet apply to you.

Free Publications About UST Requirements

Download or read *Release Detection For Underground Storage Tanks And Piping: Straight Talk On Tanks* on EPA's underground storage tank (UST) website at www.epa.gov/ust. Order printed copies of many, but not all, of our documents from the National Service Center for Environmental Publications (NSCEP), EPA's publication distributor: write to NSCEP, Box 42419, Cincinnati, OH 45242; call NSCEP's toll-free number 800-490-9198; or fax your order to NSCEP 301-604-3408.

Contents

Do You Have Questions About Release Detection? 1

An Overview Of Release Detection Requirements 2

Secondary Containment With Interstitial Monitoring 7

Automatic Tank Gauging Systems 10

Continuous In-Tank Leak Detection 13

Statistical Inventory Reconciliation 15

Tank Tightness Testing With Inventory Control 19

Manual Tank Gauging 24

Groundwater Monitoring 27

Vapor Monitoring 30

Release Detection For Underground Piping 33

Links For More Information 37

Disclaimer

This document provides information about the 2015 federal underground storage tank (UST) system requirements. The document is not a substitute for U.S. Environmental Protection Agency regulations nor is it a regulation itself — it does not impose legally binding requirements.

For regulatory requirements regarding UST systems, refer to the federal regulation governing UST systems (40 CFR part 280).

Do You Have Questions About Release Detection?

As an owner or operator of underground storage tanks (USTs) storing petroleum:

- Do you understand the basic release detection requirements for USTs?
- Do you need help choosing the best release detection method for your USTs?

These are important questions, because your UST and its underground piping must have release detection in order to comply with federal law.

This booklet begins with an overview of the federal regulatory requirements for release detection. Your implementing agency may have additional regulations, which apply to your system. Check your implementing agency requirements to ensure you are in compliance.

Throughout this document, bold type and orange updated boxes indicate new requirements in the 2015 UST regulation.

Each following section focuses on one release detection method for tanks or the requirements for piping. You will find answers in this booklet to many basic questions about how release detection methods work and which methods are best for your UST site.

Why Is Release Detection Important?

As of September 2015, over 528,000 UST releases were confirmed since the UST program was implemented. At sites without release detection, contamination can spread undetected, requiring difficult and costly cleanups.

If you have effective release detection, you can respond quickly to signs of releases. You can minimize the extent of or eliminate potential for environmental damage and the threat to human health and safety. Early action also protects you from high costs that can result from cleaning up extensive releases and responding to third-party liability claims.

State or local regulations may differ from the federal requirements. Contact your implementing agency at www.epa.gov/ust/underground-storage-tank-ust-contacts#states.

If your USTs do not meet the release detection requirements described in this booklet, you can be cited for violations and fined.

For an overview of all the federal UST requirements, see EPA's *Musts For USTs*. You can download a copy at www.epa.gov/ust.

An Overview Of Release Detection Requirements

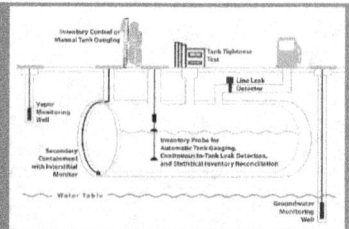

All federally regulated USTs must have a release detection method, or combination of methods, that:

- Can detect a release from any portion of the tank and the connected underground piping that routinely contains product, and
- Is installed and calibrated according to the manufacturer's instructions.

UPDATED **Tanks and piping installed or replaced after April 11, 2016 must be secondarily contained and use interstitial monitoring, except for suction piping that meets requirements discussed on page 33.**

All UST owners and operators must monitor their tanks and piping at least once every 30 days. This booklet may use the terms monthly or month and annually or annual. These terms mean at least once every 30 days and not to exceed 365 days, respectively.

For tanks installed on or before April 11, 2016, you can use any of these release detection methods:

- Secondary containment with interstitial monitoring
- Automatic tank gauging systems (performing in-tank static tests)
- Continuous in-tank leak detection
- Statistical inventory reconciliation
- Tank tightness testing with inventory control
- Manual tank gauging
- Groundwater monitoring
- Vapor monitoring
- Other methods meeting performance standards or approved by implementing agency

For underground piping installed on or before April 11, 2016, you may use any of the release detection methods listed above that are appropriate for piping or conduct periodic line tightness testing. See page 33 for piping release detection requirements.

All pressurized underground piping connected to your USTs must also have automatic line leak detectors.

For Owners Of Field-constructed Tanks Or Airport Hydrant Systems

The 2015 UST regulation removes the deferral for field-constructed tanks and airport hydrant systems, making them subject to the UST requirements. These systems are not covered in this booklet due to their uniqueness. For information on the requirements for field-constructed tanks and airport hydrant systems, see EPA's website at www.epa.gov/ust/field-constructed-tanks-and-airport-hydrant-systems-2015-requirements.

UPDATED UST systems that store fuel solely for use by emergency power generators must meet release detection requirements as follows:

- Systems installed on or before October 13, 2015 must use any of the applicable release detection methods listed above no later than October 13, 2018.
- Systems installed between October 13, 2015 and April 11, 2016 must use any of the applicable release detection methods listed above beginning at installation.
- Systems installed or replaced after April 11, 2016 must meet secondary containment requirements with interstitial monitoring as release detection.

UPDATED To make sure your release detection equipment is working properly, you must begin doing the following no later than October 13, 2018:

- Test your release detection equipment annually.
- Conduct walkthrough inspections every 30 days to visually check your release detection equipment and maintain applicable records of those checks.
- Conduct annual walkthrough inspections to visually check containment sumps and hand-held release detection equipment, such as tank gauge sticks and groundwater bailers.

UPDATED EPA revised the definition of release detection in the 2015 UST regulation. The definition clarifies that regulated substances entering into the interstitial space are leaks instead of releases. According to the 2015 UST regulation, a release always reaches the environment.

UPDATED The revised definition allows continued use of the term release detection as it applies to both releases and leaks. More importantly, the 2015 secondary containment with interstitial monitoring requirement makes it necessary to clarify how the terms release and leak are used, because product escaping the primary containment may not necessarily reach the environment.

Releases and leaks have different investigation and reporting requirements. For information on addressing suspected releases, see EPA's *Musts for USTs* at www.epa.gov/ust/musts-usts.

Look For Proof That Performance Requirements Are Met

The federal UST regulation requires that your release detection equipment meet specific performance requirements. Performance

EPA's Operating and Maintaining Underground Storage Tank Systems at www.epa.gov/ust/operating-and-maintaining-underground-storage-tank-systems-practical-help-and-checklists provides additional information about operation and maintenance.

Release means any spilling, leaking, emitting, discharging, escaping, leaching or disposing from an UST into groundwater, surface water, or subsurface soils.

Release detection means determining whether a release of a regulated substance has occurred from the UST system into the environment or a leak has occurred into the interstitial space between the UST system and its secondary barrier or secondary containment around it.

claims and means by which performance was determined must be described in writing by either the equipment manufacturer or installer. At the request of equipment manufacturers, most release detection equipment and methods available in the United States have been evaluated by a third party, who is independent of the manufacturer or vendor of the release detection system. The evaluation shows that a release detection system can work as designed. Evaluations follow recommended evaluation procedures and testing and often take place at a testing facility. EPA and third parties developed evaluation procedures for all release detection methods.

Although not mandated by federal UST requirements, many implementing agencies prefer, and some require, third party evaluation of release detection equipment and methods. Check with your implementing agency to determine what is acceptable. Although an evaluation and its resulting documentation are technical, you should be familiar with the evaluation's report and its results form. You may obtain this documentation from the release detection manufacturer and should keep it on file. Whether by the manufacturer, installer, or third party evaluation, performance claims determinations contain a signed certification that the system performed as described, as well as documentation of proper monitoring or testing procedures and any limitations of the system. This information is important to your compliance with UST requirements. For example, if a tank tightness test was evaluated and you have documentation only for tests taking two hours or more, then your UST must be tested for at least two hours or it would fail to meet the release detection requirements.

The National Work Group on Leak Detection Evaluations (NWGLDE) – an independent group – maintains a list of release detection equipment whose third-party-conducted documentation has been reviewed by the group. The list contains a detailed summary of specifications for over 390 release detection systems. Although you can use the list to help select systems and determine their compliance or acceptability, it does not consist of approved release detection systems. Approval or acceptance of release detection systems rests with your implementing agency, which in most cases is your state environmental agency. See NWGLDE's list at www.nwglde.org.

Required Probabilities For Certain Release Detection Methods

The federal UST regulation requires that release detection methods be able to detect certain leak rates consistently. Methods must detect the specified leak rate with a probability of detection of at least 95 percent and a probability of false alarm of no more than 5 percent. This means that, of 100 tests of USTs leaking at the

> *You may use any technology, as long as it meets a performance standard of detecting a leak of 0.2 gallons per hour with a probability of detection of at least 95 percent and a probability of false alarm of no more than 5 percent. Implementing agencies can approve another method if you demonstrate that it works as well as one of the methods listed in this booklet and you comply with any condition the agency imposes.*

> *Perform release detection according to documented procedures.*

specified rate, at least 95 of them must be correctly detected. It also means that, of 100 tests of non-leaking USTs, no more than 5 can be incorrectly called leaking.

Keep Release Detection Records

For each release detection method you use, you must keep these written records:

- Proof that performance claims are met and the means by which performance was determined by either the equipment manufacturer or installer and probabilities of detection and false alarm are met. Retain these records for five years or another period determined by your implementing agency.
- Results of any sampling, testing, or monitoring. Retain these results for one year or another period determined by the implementing agency. Retain tank tightness test results until the next test is conducted.
- All calibration, maintenance, and repair of release detection equipment permanently located on-site. Retain records for one year after servicing work is completed or another period determined by your implementing agency.
- Schedules of required calibration and maintenance provided by equipment manufacturers. Retain the schedules for five years from the date of installation.
- **UPDATED** **Other records may be required and are discussed, as applicable, for individual release detection methods.**

Keep Records Demonstrating Compatibility

UPDATED **The 2015 UST regulation includes additional requirements to help owners and operators demonstrate that each UST system is compatible with certain regulated substances before storing them. If you store regulated substances containing greater than 10 percent ethanol or greater than 20 percent biodiesel, or any other regulated substance identified by your implementing agency, you must keep records demonstrating compatibility of the UST system, including release detection equipment, for as long as the UST system stores the regulated substance. For more information on compatibility requirements, see EPA's *UST System Compatibility With Biofuels* at** www.epa.gov/ust/ust-system-compatibility-biofuels.

Responding To Alarms And Other Suspected Releases

Alarms associated with release detection monitoring may indicate a release has occurred. An alarm incident does not necessarily have to be reported. In the event of an alarm, you must investigate,

> *Not all release detection methods must meet required probabilities. The requirement applies to all tank release detection methods except for secondary containment with interstitial monitoring and groundwater and vapor monitoring. It also applies to automatic line leak detectors and line tightness testing.*

> *Make sure your UST system is compatible with the substance it stores.*

determine, and correct the source of the alarm. Suspected releases must be reported to your implementing agency within 24 hours or another period specified by your implementing agency. Check with your implementing agency to determine whether the alarm incident must also be reported.

Secondary Containment With Interstitial Monitoring

Secondary containment uses a barrier, an outer wall, or a liner around the UST or piping to provide secondary containment. Tanks can also be equipped with inner bladders that provide secondary containment.

UPDATED **Tanks and piping installed or replaced after April 11, 2016 must be secondarily contained and use interstitial monitoring. This applies to UST systems containing petroleum or hazardous substances.**

Will You Be In Compliance?

When installed and operated according to the manufacturer's specifications, secondary containment with interstitial monitoring meets the federal release detection requirements for USTs. You must test for a release at least once every 30 days. Secondary containment with interstitial monitoring can also be used to detect leaks from piping. See release detection for piping requirements on page 33.

How Does The Release Detection Method Work?

Secondary containment provides a barrier between the tank and the environment. The barrier holds the leak between the tank and the barrier so that the leak is detected. The barrier is shaped so that a leak will be directed toward the interstitial monitor. Barriers include:

- Double-walled or jacketed tanks, in which an outer wall partially or completely surrounds the primary tank;
- Internally fitted liners, such as bladders; and
- Leak proof excavation liners that partially or completely surround the tank.

Clay and other earthen materials are not considered acceptable secondary barriers.

Monitors are used to check the area between the tank and the barrier for leaks and alert the operator if a leak is suspected.

Some monitors indicate the physical presence of the leaked product, either liquid or gaseous. Other monitors check for a change in condition that indicates a hole in the tank, such as a

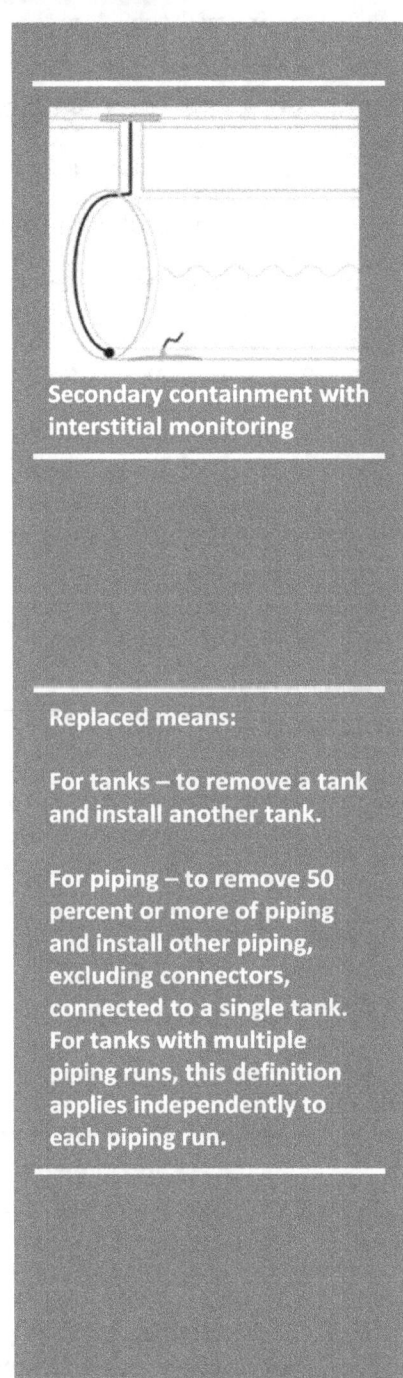

Secondary containment with interstitial monitoring

Replaced means:

For tanks – to remove a tank and install another tank.

For piping – to remove 50 percent or more of piping and install other piping, excluding connectors, connected to a single tank. For tanks with multiple piping runs, this definition applies independently to each piping run.

loss of vacuum or pressure, or a change in the level of a monitoring liquid, such as a brine or glycol solution, between the walls of a double-walled tank.

Monitors can be as simple as a dipstick used at the lowest point of the containment to see if liquid product has leaked and pooled there. Monitors can also be sophisticated automated systems that continuously check for leaks.

What Are The Regulatory Requirements?

You must check for a release at least once every 30 days.

The barrier must be immediately around or beneath the tank.

A double-walled system must be able to detect a leak through the inner wall.

An excavation liner must:

- Direct a leak toward the monitor;
- Prohibit the specific product stored to pass through it no faster than 10^{-6} centimeters per second;
- Be compatible with the product stored in the tank;
- Allow the UST's cathodic protection to work unaffected;
- Withstand moisture;
- Always be above the groundwater and the 25-year flood plain; and
- Have clearly marked and secured monitoring wells, if they are used.

A bladder must be compatible with the product stored and must be equipped with an automatic monitoring device.

UPDATED **No later than October 13, 2018, you must begin performing the following on your release detection equipment annually to make sure it is working properly.**

For hand held non-electronic equipment (including dipsticks):

- **Check for operability and serviceability**
- **Keep walkthrough inspection records for one year**

For other equipment:

- **Verify the system configuration of the controller**
- **Test alarm operability and battery backup**
- **Inspect sensors for residual build-up**
- **Ensure sensor communication with controller**
- **Keep records of these tests for three years**

These activities must be performed according to manufacturer's requirements; a nationally recognized code of practice; or requirements determined by your implementing

agency to be as protective of human health and the environment.

An unexplained presence of liquid in the interstitial space of secondarily contained systems is considered an unusual operating condition. Except if the liquid in the interstitial space is used as part of the interstitial monitoring method, for example brine, if you find liquid in the interstitial space of secondarily contained systems, you must investigate, remove the liquid, and correct the source of the liquid.

> *You must investigate and remove any liquid in the interstitial space of secondarily contained systems, unless the liquid is part of the release detection method.*

Anything Else You Should Consider?

In areas with high groundwater or a lot of rainfall, it may be necessary to select a secondary containment system that completely surrounds the tank to prevent moisture from interfering with the monitor.

This method works effectively only if the barrier and the interstitial monitor are installed correctly. Therefore, trained and experienced installers are necessary.

Automatic Tank Gauging Systems

UPDATED In an automatic tank gauging (ATG) system, a probe permanently installed in the tank is connected to a monitor to provide information on product level and temperature. These systems calculate changes in product volume that can indicate a leaking tank. ATG systems operate in one of two modes: inventory mode and leak detection mode. In the leak detection mode, ATG systems can be set to perform a leak test on either a periodic basis or continuous basis. Leak tests set to run on a periodic basis are referred to as in-tank static tests and require the system to be taken off-line typically for between one to six hours. Leak testing set to run on a continuous basis is referred to as continuous in-tank leak detection and operates on an uninterrupted or nearly uninterrupted manner.

Will You Be In Compliance?

When installed and operated according to the manufacturer's specifications, ATG systems meet the federal release detection requirements for tanks installed on or before April 11, 2016. A leak test performed at least every 30 days is required for the tank. This method does not detect piping leaks. For piping, see release detection requirements for piping on page 33.

How Does The Release Detection Method Work?

In the inventory mode:

- The product level and temperature in a tank are measured and recorded by a computer.
- ATG systems replace the use of the gauge stick to measure product level and perform inventory control. This mode records the activities of an in-service tank, including deliveries.

UPDATED **In the leak detection mode for in-tank static testing:**

- **The tank is taken out of service and the product level and temperature are measured for at least one hour.**

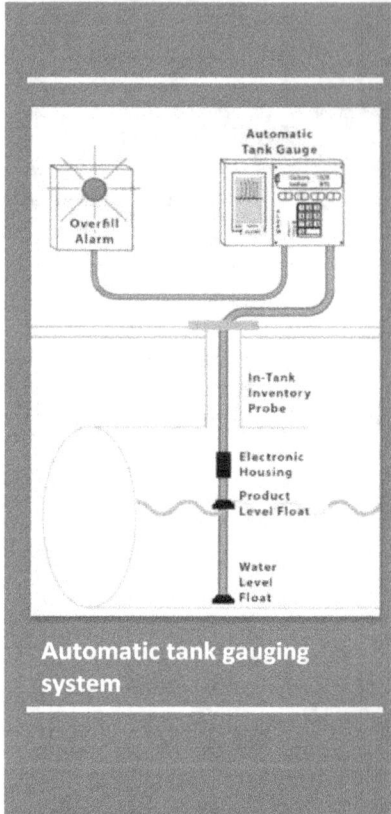

Automatic tank gauging system

In the leak detection mode, ATG systems can be set to perform either a:
- periodic leak test, also known as an in-tank static test, or
- continuous leak test, also known as continuous in-tank leak detection.

Note: When referring to ATG systems in this booklet, we mean a system performing in-tank static testing while operating in the leak detection mode. See continuous in-tank leak detection on page 13 for ATG systems performing continuous in-tank leak detection testing while operating in the leak detection mode.

In the leak detection mode for continuous in-tank leak detection:

- Some systems, known as continuous ATG systems, do not require the tank be taken out of service to perform a test. This is because these systems can gather and analyze data during many short periods when no product is being added to or taken from the tank.
- Other systems combine aspects of automatic tank gauges with statistical inventory reconciliation.

Note: Both of these methods fall under continuous in-tank leak detection because they operate on an uninterrupted basis or pause for milliseconds to gather and record data for continual analysis of the tank's leak status. See page 13 for more information about these methods.

What Are The Regulatory Requirements?

ATG systems must be able to detect a leak of at least 0.2 gallon per hour with a probability of detection of at least 95 percent and a probability of false alarm of no more than 5 percent. Some ATG systems can also detect a leak of 0.1 gallon per hour with the probabilities listed above.

UPDATED No later than October 13, 2018, you must begin performing the following on your release detection equipment annually to make sure it is working properly:

- Verify the system configuration
- Test alarm operability and battery backup
- Inspect probes and sensors for residual build-up
- Ensure floats move freely, the shaft is not damaged, and cables are free of kinks and breaks
- Keep records of these tests for three years

These activities must be performed according to manufacturer's requirements; a nationally recognized code of practice; or requirements determined by your implementing agency to be as protective of human health and the environment.

An unexplained presence of water in the tank is considered an unusual operating condition. If you find water in your tank, you

You must obtain a conclusive pass or fail result within the 30 day monitoring period. If the test report is inconclusive, you must use another method of release detection for that 30 day monitoring period. An inconclusive result means you have not performed release detection for that 30 day period.

must investigate and correct the source of the water. Suspected releases must be reported to your implementing agency within 24 hours or another period specified by your implementing agency.

Anything Else You Should Consider?

Detecting water in the tank is important. Water around a tank may mask a hole in the tank or distort the data to be analyzed by temporarily preventing a release. To detect a release in this situation, check for water at least once a month. **Depending on the product in the tank, detecting water may be difficult, but it is possible to do. Products such as ethanol-based fuels may not form a water bottom.**

ATG systems have been used primarily on tanks containing gasoline or diesel. If considering using an ATG system for larger tanks or products other than gasoline or diesel, discuss its applicability with the equipment manufacturer or installer. Check the method's documentation to confirm that it will meet regulatory requirements and your needs.

With the exception of some continuous ATG systems evaluated to perform on manifolded tanks, each tank at a site must be equipped with a separate probe. Check the method's documentation to determine if the ATG system can be used with manifolded tanks. For more information, see continuous in-tank leak detection requirements on page 13. The ATG system probe is connected to a console that displays product level information and the results of the monthly test. Printers can be connected to the console to record this information.

> *The ATG system probe is installed through an opening, which is different than the fill pipe, on the top of the tank.*

ATG systems are often equipped with alarms for high and low product level and high water level.

For ATG systems used for static release detection testing, no product can be delivered to the tank or withdrawn from it for one to six hours before the monthly test or during the test, which generally takes one to six hours. These times vary depending on the specific ATG system model. Check with your equipment manufacturer or installer. You may also find information on your ATG system on NWGLDE's list of release detection evaluations at www.nwglde.org.

> *ATG systems can be linked with computers at remote locations, from which the system can be programmed or read.*

An ATG system can be programmed to perform a test more often than once every 30 days. EPA recommends this practice.

Some ATG systems may be evaluated to test at relatively low capacities, for example, 25 percent or 30 percent. Although the product level at such capacities may be valid for the test equipment, it may not appropriately test all portions of the tank that routinely contain product. The ATG leak test must be run and tank tested at the capacity to which it is routinely filled.

Continuous In-Tank Leak Detection

The 2015 federal UST regulation added continuous in-tank leak detection (CITLD) as a release detection method and establishes requirements for its operation and maintenance. CITLD encompasses all statistically based methods where the system incrementally gathers measurements on an uninterrupted or nearly uninterrupted basis to determine a tank's leak status.

Will You Be In Compliance?

You can use CITLD methods for tanks installed on or before April 11, 2016. The system incrementally gathers measurements to determine a tank's leak status within the 30-day monitoring period. Some methods address pipelines and have been verified to meet pipeline performance standards. These methods are capable of meeting the pipeline release detection requirements. See release detection requirements for piping on page 33.

How Does The Release Detection Method Work?

There are two major groups that fit into this category: continuous statistical release detection, also referred to as continuous automatic tank gauging methods, and continual reconciliation. Both groups typically use sensors permanently installed in the tank to obtain inventory measurements. They are combined with a microprocessor in the ATG system or other control console that processes the data. Continual reconciliation methods are further distinguished by their connection to dispensing meters that allow for automatic recording and use of dispensing data in analyzing tanks' leak status.

What Are The Regulatory Requirements?

CITLD operates on an uninterrupted basis or operates by allowing the system to gather incremental measurements to determine the release status of the tank at least once every 30 days.

Continuous in-tank leak detection

CITLD must be able to detect a leak at least 0.2 gallon per hour with a probability of detection of at least 95 percent and a probability of false alarm of no more than 5 percent. Some CITLD methods can also detect a leak of 0.1 gallon per hour with the probabilities listed above.

UPDATED **No later than October 13, 2018, you must begin performing the following on your release detection equipment annually to make sure it is working properly:**

- **Verify the system configuration of the controller**
- **Test alarm operability and battery backup**
- **Inspect probes and sensors for residual build-up**
- **Ensure floats move freely, the shaft is not damaged, and cables are free of kinks and breaks**
- **Keep records of these tests for three years**

These activities must be performed according to manufacturer's instructions; a nationally recognized code of practice; or requirements determined by your implementing agency to be as protective of human health and the environment.

An unexplained presence of water in the tank is considered an unusual operating condition. If you find water in your tank you must investigate and correct the source of the water. Suspected releases must be reported to your implementing agency within 24 hours or another period specified by your implementing agency.

Anything Else You Should Consider?

Detecting water in the tank is important. Water around a tank may mask a hole in the tank or distort the data to be analyzed by temporarily preventing a release. To detect a release in this situation, check for water at least once a month. **Depending on the product in the tank, detecting water may be difficult, but it**
UPDATED **is possible to do. Products such as ethanol-based fuels may not form a water bottom.**

> *You must obtain a conclusive pass or fail result within the 30 day monitoring period. If the test report is inconclusive, you must use another method of release detection for that 30 day monitoring period. An inconclusive result means you have not performed release detection for that 30 day period.*

> *See NWGLDE at www.nwglde.org, which is a source for checking whether your CITLD method meets regulatory performance requirements.*

> *CITLD methods may allow for monitoring larger tank capacities and higher system throughputs. However, these methods have limitations as well.*

Statistical Inventory Reconciliation

The 2015 federal UST regulation added statistical inventory reconciliation (SIR) as a release detection method. For this method, a trained professional uses sophisticated computer software to conduct a statistical analysis of inventory, delivery, and dispensing data, which is gathered periodically and supplied regularly to the vendor.

Will You Be In Compliance?

SIR, when performed according to the vendor's specifications, meets federal release detection requirements for USTs and piping installed on or before April 11, 2016. SIR with a 0.2 gallon per hour release detection capability meets the federal requirements for monthly monitoring for tanks. SIR with a 0.1 gallon per hour release detection capability meets the federal requirements as an equivalent to tank tightness testing. If it has the capability of detecting even smaller leaks, SIR meets the federal requirements for line tightness testing as well. See release detection requirements for piping on page 33.

How Does The Release Detection Method Work?

SIR analyzes inventory, delivery, and dispensing data collected over a period of time to determine whether or not a tank or piping is leaking a regulated substance.

Each operating day, the product level is measured using a gauge stick or other tank level monitor. You must also keep complete records of all withdrawals from the UST and all deliveries to the UST. After data have been collected for the period of time required by the SIR vendor, you provide the data to the SIR vendor.

The SIR vendor conducts a statistical analysis of the data to determine whether or not your UST system is leaking. The SIR vendor provides you with a test report of the analysis. Alternatively, you can purchase SIR software, which performs this same analysis and provides a test report from your own computer.

Some methods combine aspects of automatic tank gauges with statistical inventory reconciliation. In these methods,

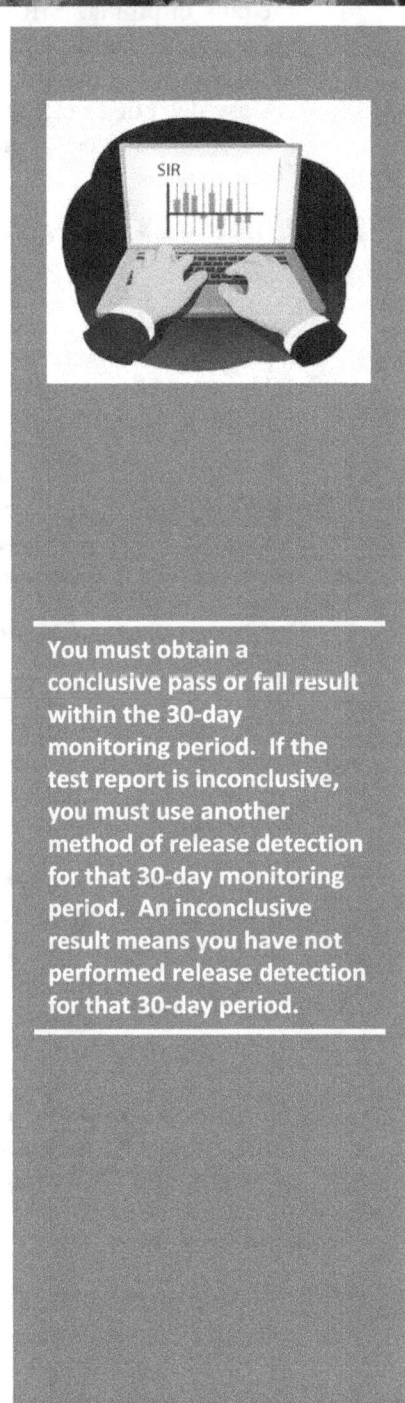

You must obtain a conclusive pass or fail result within the 30-day monitoring period. If the test report is inconclusive, you must use another method of release detection for that 30-day monitoring period. An inconclusive result means you have not performed release detection for that 30-day period.

sometimes called hybrid methods, a gauge provides liquid level and temperature data to a computer running SIR software, which performs the analysis to detect leaks.

SIR methods are distinguished from continuous in-tank leak detection methods by how inventory, delivery, and dispensing data are processed; they provide a determination of the release status of the tank or piping. SIR data is processed on a periodic basis involving a separate analysis that is performed either by a SIR vendor or SIR software. Continuous statistically based in-tank release detection methods process data on an on-going, uninterrupted or nearly uninterrupted manner.

What Are The Regulatory Requirements?

SIR methods must report a quantitative result with a calculated leak rate, be able to detect a leak at least 0.2 gallons per hour with a probability of detection of at least 95 percent and a probability of false alarm of no more than 5 percent. Some SIR methods can also detect a leak of 0.1 gallons per hour with the probabilities listed above.

UPDATED **No later than October 13, 2018, you must begin performing the following on your release detection equipment annually to make sure it is working properly:**

For hand held non-electronic equipment, such as tank gauge sticks:

- **Check for operability and serviceability**
- **Keep walkthrough inspection records for one year**

For other equipment:

- **Verify the system configuration of the controller**
- **Test alarm operability and battery backup**
- **Inspect probes and sensors for residual build-up**
- **Ensure floats move freely, the shaft is not damaged, and cables are free of kinks and breaks**
- **Keep records of these tests for three years**

These activities must be performed according to manufacturer's instructions; a nationally recognized code of practice; or requirements determined by your implementing agency to be as protective of human health and the environment.

UPDATED **The SIR method must use a threshold that does not exceed one-half the minimum detectable leak rate (MDL). Pd is the probability of detection and Pfa is the probability of false alarm in a normal probability distribution. SIR data is typically analyzed through the calculation of the reportable**

EPA's Introduction To Statistical Inventory Reconciliation For Underground Storage Tanks at www.epa.gov/ust/introduction-statistical-inventory-reconciliation-underground-storage-tanks explains how to do statistical inventory reconciliation.

values of MDL and the leak declaration threshold T are related as follows:

- MDL is always greater than T
- Pd = (1 – Pfa), then MDL = 2 times T (that is, the threshold is equal to ½ MDL)

Any analysis of data indicating a threshold value greater than one-half MDL should be appropriately investigated as a suspected release.

You must keep on file for one year the test reports. You must also keep on file for five years documentation that the SIR method used for your system is capable of detecting a leak rate of 0.2 gallons per hour with a probability of detection of 95 percent and probability of false alarm.

Documentation on the method's capability of meeting performance requirements must reflect the way the method is used in the field.

An unexplained presence of water in the tank is considered an unusual operating condition. If you find water in your tank you must investigate and correct the source of the water. Suspected releases must be reported to your implementing agency within 24 hours or another period specified by your implementing agency.

Anything Else You Should Consider?

Detecting water in the tank is important. Water around a tank may mask a hole in the tank or distort the data to be analyzed by temporarily preventing a release. To detect a release in this situation, check for water at least once a month. **Depending on the product in the tank, detecting water may be difficult, but it is possible to do. Products such as ethanol-based fuels may not form a water bottom.**

If you are considering using a SIR method, check the method's documentation to confirm that it will meet regulatory requirements and your specific UST system needs.

A SIR method's ability to detect releases declines as throughput increases. If you are considering using a SIR method for high throughput UST systems, check the method's documentation to confirm that it will meet regulatory requirements and your needs.

Data, including product level measurements, dispensing data, and delivery data, should all be carefully collected according to the SIR vendor's specifications. Poor data collection can produce inconclusive results and noncompliance.

The SIR vendor will generally provide forms for recording data, a calibrated chart converting liquid level to volume, and detailed instructions on conducting measurements.

UPDATED

SIR should not be confused with other release detection methods that also rely on periodic reconciliation of inventory, withdrawal, and delivery data. Unlike manual tank gauging or inventory control, SIR uses a sophisticated statistical analysis of data to detect releases.

Tank Tightness Testing With Inventory Control

This method combines periodic tank tightness testing with monthly inventory control. Inventory control involves taking measurements of tank contents and recording amount received and pumped each operating day, as well as reconciling all this data at least once every 30 days. Every five years, this combined method must also include a tightness test, which is a sophisticated test performed by a trained professional.

Will You Be In Compliance?

When performed according to the manufacturer's specifications, periodic tank tightness testing combined with monthly inventory control can temporarily meet the federal release detection requirements for tanks installed on or before April 11, 2016. This method does not detect piping leaks. This combined method can be used only for 10 years after the tank was installed.

These two release detection methods must be used together because inventory control alone does not meet the federal requirements for monthly release detection for tanks. Line tightness testing, a separate type of tightness testing, is also an option for underground piping; see release detection requirements for piping on page 33.

We discuss both tank tightness testing and inventory control below. We discuss tank tightness testing first, followed by inventory control. Tank tightness testing is also used in combination with manual tank gauging as described on page 24. In addition, tank tightness testing may be used to investigate a suspected release.

Tank tightness testing

Tank Tightness Testing

How Does The Release Detection Method Work?

Tightness tests, also referred to as precision tank tests, include a variety of methods. These methods are divided into two categories: volumetric and nonvolumetric.

Volumetric test methods generally involve precisely measuring in milliliters or thousandths of an inch the change in product level in a tank over time. Additional characteristics of this category of tank tightness testing include:

- Changes in product temperature also must be precisely measured in thousandths of a degree at the same time as level measurements, because temperature changes cause volume changes that interfere with finding a leak.
- The product in the tank must be at a certain minimum level before testing. This often requires adding product from another tank on site or purchasing additional product.
- A net decrease in product volume, which you find by subtracting out volume changes caused by temperature, over the time of the test indicates a leak.
- A few of these methods measure properties of product that are independent of temperature, such as mass, and so do not need to measure product temperature.

There are many nonvolumetric test methods. These methods can be distinguished by what they test or which areas of the UST system they test. The methods:

- Involve acoustics that interpret an ultrasonic signal.
- Use vacuum or pressure decay with gain or loss of pressure, respectively, to determine whether there is a hole in the tank.
- Test either the wetted portion of the tank, which contains product, or the ullage, which is the unfilled portion of the tank.
- Involve tracer compounds circulated through the UST system, which test strategically placed sampling ports outside the UST system.

Although not typically done, you may use tank tightness testing to meet the monthly release detection requirement. This test must meet performance standards of 0.1 gallon per hour leak rate with probability of detection at least 95 percent and probability of false alarm not to exceed 5 percent.

Except for tracer compounds used for both volumetric and nonvolumetric test methods, the following generally apply:

- The testing equipment is temporarily installed in the tank, usually through the fill pipe.
- The tank must be taken out of service for the test.
- Some tightness test methods require the tester measure and calculate by hand. Other tightness test methods are highly

automated. After the tester sets up the equipment, a computer controls the measurements and analysis.

- Some ATG systems are capable of meeting the regulatory requirements for tank tightness testing and may be considered an equivalent method. Check with your implementing agency.

What Are The Regulatory Requirements?

The tightness test method must be able to detect a leak at least 0.1 gallon per hour with a probability of detection of at least 95 percent and a probability of false alarm of no more than 5 percent. **No later than October 13, 2018, you must begin testing your release detection equipment annually to make sure it is working properly.**

Tank tightness testing is typically performed by a qualified testing company. Qualified testing companies periodically calibrate and maintain their equipment according to applicable standards. If your implementing agency allows use of ATG systems or other system controllers for tank tightness testing, you must follow the testing procedures required for ATG systems. See page 10.

You must perform a tightness test at least every 5 years. You may use this combination method temporarily for up to 10 years after the UST was installed. After 10 years, you must use a different release detection method.

Anything Else You Should Consider?

For most methods, a testing company performs the test. You should observe the test.

Under the federal UST regulation, this combination method can only be used for 10 years after the tank was installed. However, most states have secondary containment with interstitial monitoring requirements. Therefore, you may not be able to use this combination method. Check with your implementing agency.

Depending on the method, tank tightness testing can be used on tanks of varying capacity and tanks containing gasoline and diesel. Many test methods have limitations on the capacity of the tank or the amount of ullage, which is the unwetted portion of the tank that should not be exceeded. Methods that use tracer chemical analysis do not have limitations on tank capacity. If you are considering using tightness testing for products other than gasoline or diesel, discuss the method's applicability with the manufacturer's representative. Check the method's documentation to confirm that it will meet regulatory requirements and your specific UST system needs.

Manifolded tanks generally should be isolated and tested separately.

Procedure and personnel, not equipment, are usually the most important factors in a successful tightness test. Therefore, well-trained and experienced testers are very important. Some implementing agencies have tester certification programs.

Inventory Control

How Does The Release Detection Method Work?

Inventory control requires frequent measurements of tank contents and math calculations that let you compare your stick inventory, which is what you measured, to your book inventory, which is what your recordkeeping indicates you should have. Some people call this process inventory reconciliation. If the difference between your stick and book inventory is too large, your tank may be leaking.

UST inventories are determined each operating day by using a gauge stick and recording the data on a form. The level on the gauge stick is converted to a volume of product in the tank using a calibration chart, which is often furnished by the UST manufacturer.

The amounts of product delivered to and withdrawn from the UST each operating day are also recorded. At least once every 30 days, the gauge stick data and the sales and delivery data are reconciled and the month's overage or shortage is determined. If the overage or shortage is greater than or equal to 1 percent of the tank's flow-through volume plus 130 gallons of product, the UST may be leaking.

What Are The Regulatory Requirements?

Inventory control must be used in combination with tank tightness testing performed at least every 5 years to meet the monthly release detection requirement. This combination method can only be used for up to 10 years after the tank was installed. This method may not be used for UST systems installed after April 11, 2016.

The gauge stick must reach the bottom of the tank and be marked so that the product level can be determined to the nearest one-eighth of an inch. A monthly measurement must be taken to identify any water in the tank.

Product dispensers must be calibrated to the applicable weights and measures standards.

UPDATED **No later than October 13, 2018, you must begin performing the following on your release detection equipment annually to make sure it is working properly.**

EPA's Doing Inventory Control Right at www.epa.gov/ust/doing-inventory-control-right-underground-storage-tanks explains how to do inventory control. The booklet also contains standard recordkeeping forms.

You may need to get a corrected tank chart if your tank is not level.

For hand held non-electronic equipment, such as tank gauge sticks:

- **Check for operability and serviceability**
- **Keep walkthrough inspection records for one year**

These activities must be performed according to manufacturer's instructions; a nationally recognized code of practice; or requirements determined by your implementing agency to be as protective of human health and the environment.

An unexplained presence of water in the tank is considered an unusual operating condition. If you find water in your tank you must investigate and correct the source of the water. Suspected releases must be reported to your implementing agency within 24 hours or another period specified by your implementing agency.

Anything Else You Should Consider?

Detecting water in the tank is important. Water around a tank may mask a hole in the tank or distort the data to be analyzed by temporarily preventing a release. To detect a release in this situation, check for water at least once a month. **Depending on the product in the tank, detecting water may be difficult, but it is possible to do. Products such as ethanol-based fuels may not form a water bottom.**

UPDATED

The accuracy of tank gauging can be increased by spreading product finding paste on the gauge stick before taking measurements or by using in tank product level monitoring devices.

Inventory control is a practical, commonly used management practice that does not require closing down the tank operation for long periods.

Manual Tank Gauging

Manual tank gauging requires keeping the tank undisturbed for at least 36-58 hours each week, during which the contents of the tank are measured twice at the beginning and twice at the end of the test period. At the end of each week, you compare the results to the standards shown on page 25 to see if your tank is leaking.

Will You Be In Compliance?

Manual tank gauging can be used only on tanks containing 2,000 gallons or less. Tanks containing 1,000 gallons or less can use this method alone, if they meet specified diameter requirements discussed below. Tanks from 1,001 to 2,000 gallons, and tanks between 551 and 1,000 gallons that do not meet the specified diameters, can temporarily use manual tank gauging when it is combined with tank tightness testing. Under the federal UST regulation, this combined method can be used only for 10 years after the tank was installed. This method may not be used for UST systems installed after April 11, 2016.

Manual tank gauging detects leaks only from tanks; this method does not detect piping leaks. For requirements for piping, see release detection requirements for piping on page 33.

How Does The Release Detection Method Work?

You must take four measurements of the tank's contents, two at the beginning and two at the end of a 36-58 hour period, during which nothing is added to or removed from the tank. See the table on page 25.

The average of the two consecutive ending measurements are subtracted from the average of the two beginning measurements to indicate the change in product volume.

Every week, you compare the calculated change in tank volume to the standards shown in the table on page 25. If the calculated change exceeds the weekly standard, the UST may be leaking. Also, you must compare the averages of the four weekly test results to the monthly standard in the same way. See the table below.

Manual tank gauging

EPA's *Manual Tank Gauging For Small Underground Storage Tanks* at www.epa.gov/ust/manual-tank-gauging-small-underground-storage-tanks explains how to do manual tank gauging correctly and contains standard recordkeeping forms.

What Are The Regulatory Requirements?

You must take liquid level measurements with a gauge stick that is marked to measure the liquid to the nearest one-eighth of an inch.

UPDATED **No later than October 13, 2018, you must begin performing the following on your release detection equipment annually to make sure it is working properly.**

For hand held non-electronic equipment, such as tank gauge sticks:

- **Check for operability and serviceability**
- **Keep walkthrough inspection records for one year**

You must perform these activities according to manufacturer's instructions; a nationally recognized code of practice; or requirements determined by your implementing agency to be as protective of human health and the environment.

Manual tank gauging may be used as the sole method of release detection for tanks with a capacity of 550 gallons or less and capacities between 551 and 1,000 gallons with a 48 inch or 64 inch diameter. All other tanks using manual tank gauging must combine the method with tank tightness testing. **These tanks may use the combined method for up to 10 years after installation. After 10 years, you must use another release detection method.** See the other sections of this booklet for allowable monthly monitoring methods.

> *Under the federal UST regulation, you may only use this combination method for 10 years after the tank was installed. However, most states have secondary containment with interstitial monitoring requirements. Therefore, you may not be able to use this combination method. Check with your UST implementing agency.*

Table Of Test Standards For Manual Tank Gauging

Tank Size	Minimum Duration Of Test	Weekly Standard (1 test)	Monthly Standard (4-test average)
Up to 550 gallons	36 hours	10 gallons	5 gallons
551-1,000 gallons (when tank diameter is 64")	44 hours	9 gallons	4 gallons
551-1,000 gallons (when tank diameter is 48")	58 hours	12 gallons	6 gallons
551-1,000 gallons (also requires periodic tank tightness testing)	36 hours	13 gallons	7 gallons
1,001-2,000 gallons (also requires periodic tank tightness testing)	36 hours	26 gallons	13 gallons

An unexplained presence of water in the tank is considered an unusual operating condition. If you find water in your tank, you must investigate and correct the source of the water. You must report suspected releases to your implementing agency within 24 hours or the period specified by your implementing agency.

Anything Else You Should Consider?

Detecting water in the tank is important. Water around a tank may mask a hole in the tank or distort the data to be analyzed by temporarily preventing a release. To detect a release in this situation, check for water at least once a month. **Depending on the product in the tank, detecting water may be difficult, but it is possible to do. Products such as ethanol-based fuels may not form a water bottom.**

UPDATED

You can perform manual tank gauging yourself. Correct gauging, recording, and correct math are the most important factors for successful tank gauging. The accuracy of manual tank gauging can be increased by spreading product-finding paste on the gauge stick before taking measurements.

Groundwater Monitoring

Groundwater monitoring detects the presence of liquid product floating on the groundwater near the tank and along the piping runs. To discover if released product has reached groundwater, these wells can be checked periodically using hand-held equipment or continuously with permanently installed equipment.

Groundwater monitoring

Will You Be In Compliance?

When installed and operated according to the manufacturer's instructions, a groundwater monitoring system can meet the federal release detection requirements for USTs and piping installed on or before April 11, 2016. Monitoring of a groundwater monitoring system is required at least once every 30 days for the tank.

UPDATED **No later than October 13, 2018, if you use groundwater monitoring, you must begin keeping records of a site assessment, for as long as you use this method, showing that the monitoring system is installed properly. Site assessments performed after October 13, 2015 must be signed by a licensed professional.**

How Does The Release Detection Method Work?

Groundwater monitoring involves the use of permanent monitoring wells placed close to the UST, with the wells extending below the groundwater level. The wells are checked at least every 30 days for the presence of product that has leaked from the UST and is floating on the groundwater.

The two main components of a groundwater monitoring system are the monitoring wells, which are typically at least 4 inches in diameter, and the monitoring device.

Electronic detection devices may be permanently installed in the well for automatic, continuous measurements for released product.

Manual devices range from a bailer, which collects a liquid sample for visual inspection, to a device that can be inserted into the well to electronically indicate the presence of leaked product. Manual devices must be used to check each monitoring well at least once every 30 days.

Before installation, a site assessment is necessary to determine the soil type, groundwater depth and flow direction, and the general geology of the site. A trained professional must perform this assessment.

The number of wells and their placement is very important. Only an experienced contractor can properly design and construct an effective monitoring well system. A minimum of two wells is recommended for a single tank excavation. Three or more wells are recommended for an excavation with two or more tanks. Some implementing agencies have developed rules for monitoring well placement.

What Are The Regulatory Requirements?

Groundwater monitoring can only be used if the stored substance does not mix with water and floats on top of water.

If groundwater monitoring is used as the sole method of release detection, the groundwater must be less than 20 feet below the surface, and the soil between the well and the UST must be sand, gravel, or other coarse materials.

Product detection devices must be able to detect one-eighth inch or less of leaked product on top of the groundwater.

Monitoring wells must be properly designed and sealed to keep them from becoming contaminated from outside sources.

Wells should be placed in the UST backfill so they can detect a leak as quickly as possible.

Monitoring wells must be secured and clearly marked.

UPDATED **No later than October 13, 2018, you must begin performing the following on your release detection equipment annually to make sure it is working properly.**

For hand held non-electronic equipment, such as groundwater bailers:

- **Check for operability and serviceability**
- **Keep walkthrough inspection records for one year**

For other equipment:

- **Verify the system configuration of the controller**
- **Test alarm operability and battery backup**
- **Inspect well probes and sensors for residual build-up**

> *No later than October 13, 2018, if you use vapor monitoring or groundwater monitoring, you must begin keeping records of a site assessment, for as long as you use these methods, showing that the monitoring system is set up properly. If you do not have a site assessment for your vapor monitoring or groundwater monitoring, you will need to have one conducted. Site assessments conducted after October 13, 2015 must be signed by a licensed professional.*

> *Groundwater at times may be more than 20 feet from the ground surface, due to seasonal water table variations. This can result in the depth to groundwater requirement not being met.*

- **Ensure floats move freely, the shaft is not damaged, and cables are free of kinks and breaks**
- **Test manual electronic devices, such as portable probes**
- **Keep records of these tests for three years**

These activities must be performed according to manufacturer's requirements; a nationally recognized code of practice; or requirements determined by your implementing agency to be as protective of human health and the environment.

Anything Else You Should Consider?

In general, groundwater monitoring works best at UST sites where:

- Monitoring wells are installed in the tank backfill; and
- There are no previous releases of product that would falsely indicate a current release.

A professionally conducted site assessment is critical for determining these site-specific conditions.

UPDATED

Some states may allow you to use groundwater monitoring wells to perform vapor monitoring. Check with your implementing agency to determine what is acceptable. If allowed, unless an analysis is performed and valid documentation regarding use of the wells for vapor monitoring during low water table conditions is identified in the site assessment, the wells will be restricted for groundwater monitoring only.

UPDATED

In the event of a confirmed release at an UST site, groundwater monitoring is no longer acceptable to meet the release detection requirement unless the site is remediated and a new site assessment is conducted.

Vapor Monitoring

Vapor monitoring measures either product vapors in the soil around the UST, referred to as passive monitoring, or special tracer chemicals added to the UST, referred to as active monitoring.

Will You Be In Compliance?

When installed and operated according to the manufacturer's instructions, vapor monitoring can meet the federal release detection requirements for tanks and piping installed on or before April 11, 2016. Monitoring of a vapor monitoring system at least every 30 days is required for the tank.

UPDATED **No later than October 13, 2018, if you use vapor monitoring you must begin keeping records of a site assessment, for as long as you use this method, showing that the monitoring system is installed properly. Site assessments performed after October 13, 2015 must be signed by a licensed professional.**

How Does The Release Detection Method Work?

Vapor monitoring can be categorized into two types: active monitoring and passive monitoring. Active monitoring is also referred to as chemical marker monitoring or as tracer compound analysis.

Passive monitoring detects or measures vapors from released product within monitoring wells placed in the soil around the tank to determine if the tank is releasing regulated substances.

Active monitoring samples for the presence of a tracer compound outside the UST system that was introduced in the tank or underground piping.

Fully automated vapor monitoring systems have permanently installed equipment to continuously or periodically gather and analyze vapor samples and activate a visual or audible alarm when a release is detected, independent of actions by an UST system operator.

Active monitoring requires the installation of monitoring wells or sampling points strategically placed in the tank

Vapor monitoring

backfill or along pipe runs to intercept special chemicals that, in the event of a release, are detected in the sampling points.

Manually operated vapor monitoring systems range from equipment that immediately analyzes a gathered vapor sample to devices that gather a sample, which must be sent to a laboratory for analysis. Manual systems must be used at least once every 30 days to monitor a site. If active monitoring is performed, it must be done at least every 30 days by qualified technicians.

Before installation of any vapor monitoring system for release detection, a site assessment is necessary to determine the soil type, groundwater depth and flow direction, and the general geology of the site. Only a trained professional can do this.

The number of wells and their placement is very important. Only an experienced contractor can properly design and construct an effective monitoring well system. Vapor monitoring requires installation of monitoring wells within the tank backfill. A minimum of two wells is recommended for a single tank excavation. Three or more wells are recommended for an excavation with two or more tanks. Some implementing agencies have developed requirements for monitoring well placement.

What Are The Regulatory Requirements?

The UST backfill must be sand, gravel, or another material that will allow petroleum vapors or tracer compound to easily move to the monitor.

The backfill must be clean enough that previous contamination does not interfere with detecting a current release.

The substance stored in the UST must vaporize easily so that the vapor monitor can detect a release. For example, some vapor monitoring systems do not work well, if at all, with diesel fuel.

High groundwater, excessive rain, or other sources of moisture must not interfere with operation of vapor monitoring for more than 30 consecutive days.

Monitoring wells must be secured and clearly marked.

UPDATED

No later than October 13, 2018, you must begin performing the following on your release detection equipment annually to make sure it is working properly.

For hand held non-electronic equipment:

- **Check for operability and serviceability**
- **Keep walkthrough inspection records for one year**

For other equipment:

> *To ensure they are properly operating, vapor monitoring devices must be periodically calibrated according to the manufacturer's instructions.*

> *No later than October 13, 2018, if you use vapor monitoring or groundwater monitoring, you must keep records of a site assessment, for as long as you use these methods, showing that the monitoring system is set up properly. If you do not have a site assessment for your vapor monitoring or groundwater monitoring, you will need to have one conducted. Site assessments conducted after October 13, 2015 must be signed by a licensed professional.*

- Verify the system configuration of the controller
- Test alarm operability and battery backup
- Inspect sensors for residual build-up
- Test manual electronic devices, such as photoionization detectors
- Keep records of these tests for three years

These activities must be performed according to manufacturer's instructions; a nationally recognized code of practice; or requirements determined by your implementing agency to be as protective of human health and the environment.

Anything Else You Should Consider?

Before installing a vapor monitoring system, a site assessment must be done to determine whether vapor monitoring is appropriate at the site. A site assessment usually includes at least a determination of the groundwater level, background contamination, stored product type, and soil type. This assessment can only be done by a trained professional.

UPDATED In the event of a confirmed release at an UST site, vapor monitoring is no longer acceptable to meet the release detection requirement unless the site is remediated and a new site assessment is conducted.

Release Detection For Underground Piping

Owners and operators of federally regulated UST systems must have a release detection method, or combination of methods for connected underground piping that routinely contains product.

Will You Be In Compliance?

When installed and operated according to the manufacturer's specifications, the release detection methods discussed here meet the federal regulatory requirements for underground piping systems. Your UST may have suction or pressurized piping, which are discussed below.

UPDATED **Piping installed or replaced after April 11, 2016 must have secondary containment with interstitial monitoring, except for suction piping that meets requirements discussed below. In addition, pressurized piping must have an automatic line leak detector.**

What Are The Regulatory Requirements For Suction Piping?

No release detection is required if the suction piping system has these characteristics: below-grade piping that operates under atmospheric pressure; enough slope so that the product in the pipe can drain back into the tank when suction is released; and only one check valve, which is located as close as possible beneath the pump in the dispensing unit. If a suction line is to be considered exempt based on these characteristics, there must be some way to verify that the line actually has these characteristics.

Suction piping installed on or before April 11, 2016 that does not have all of the characteristics noted above must use one of the following release detection methods:

- A line tightness test at least every three years
- Monthly interstitial monitoring
- Monthly vapor monitoring
- Monthly groundwater monitoring
- Monthly statistical inventory reconciliation

Line leak detection

- Continuous in-tank leak detection only for methods that include pipelines
- Other monthly monitoring that meets performance standards or approved by your implementing agency

Suction lines are not pressurized very much during a tightness test of 7 to 15 pounds per square inch.

Interstitial monitoring, vapor monitoring, groundwater monitoring, continuous in-tank leak detection, and statistical inventory reconciliation have the same regulatory requirements for piping as they do for tanks. See earlier sections of this booklet for information on those methods.

UPDATED **Suction piping installed or replaced after April 11, 2016 that does not meet all of the design standards above must have secondary containment with interstitial monitoring.**

What Are The Regulatory Requirements For Pressurized Piping?

Pressurized piping installed on or before April 11, 2016 must have an automatic line leak detector (ALLD) that:

- Shuts off flow, or
- Restricts flow, or
- Triggers an audible or visual alarm

The ALLD must be designed to detect a release at least 3 gallons per hour at a line pressure of 10 pounds per square inch within 1 hour, with a probability of detection of at least 95 percent and a probability of false alarm of no more than 5 percent.

You must also use one of these other methods:

- Annual line tightness test
- Monthly interstitial monitoring
- Monthly vapor monitoring
- Monthly groundwater monitoring
- Monthly statistical inventory reconciliation
- **Continuous in-tank leak detection, only for methods that include pipelines**
- Other monthly monitoring that meets performance standards or approved by your implementing agency

UPDATED

The line tightness test must be able to detect a leak at least 0.1 gallon per hour with a probability of detection of at least 95 percent and a probability of false alarm of no more than 5 percent when the line pressure is 1.5 times its normal operating pressure. The test must be conducted each year. If the test is performed at pressures lower than 1.5 times operating pressure, the leak rate to be detected must be correspondingly lower.

Interstitial monitoring, vapor monitoring, groundwater monitoring, continuous in-tank leak detection only for methods that include piping, and statistical inventory reconciliation have the same regulatory requirements for piping as for tanks. See earlier sections of this booklet for information on those methods.

UPDATED **Pressurized piping installed or replaced after April 11, 2016 must have secondary containment with interstitial monitoring.**

UPDATED **No later than October 13, 2018, you must begin annual operability testing of ALLDs to determine they are capable of detecting a leak of 3 gallons per hour at 10 pounds per square inch line pressure within 1 hour by simulating a leak at or below this leak rate. You must keep records of these tests for 3 years.**

The test must be performed according to manufacturer's instructions; a nationally recognized code of practice; or requirements determined by your implementing agency to be as protective of human health and the environment.

If your implementing agency allows an ALLD to meet other aspects of the pressurized piping dual release detection requirement (that is, monthly monitoring or line tightness testing), the annual operability test must be conducted to ensure the applicable performance standard can be met. Simulating a leak at 0.2 gph for monthly monitoring or 0.1 gph for line tightness testing is one way to ensure this.

How Do The Release Detection Methods Work?

ALLDs

Flow restrictors and flow shutoffs can monitor the pressure within the line in a variety of ways: whether the pressure decreases over time; how long it takes for a line to reach operating pressure; and combinations of increases and decreases in pressure.

All mechanical and electronic ALLDs must meet the annual testing requirement.

If a suspected release is detected, a flow restrictor keeps the product flow through the line well below the usual flow rate. If a suspected release is detected, a flow shutoff completely cuts off product flow in the line or shuts down the pump.

Both automatic flow restrictors and shutoffs are permanently installed directly into the pipe or the pump housing.

A continuous alarm system constantly monitors line conditions and immediately triggers an audible or visual alarm if a release is suspected. An automated interstitial monitoring system can be set to operate continuously independent of an operator and sound an alarm, flash a signal on the console, or even ring a telephone in a manager's office when a release is suspected.

A self diagnostic system does not meet the annual testing requirement, unless the system performs a simulated leak test.

UPDATED **An automated interstitial monitoring system can be combined with an automatic shutoff system so that whenever the system detects a suspected release, the product flow in the piping is completely shut down. Under other methods in 40 CFR §**

280.43(i)(2), EPA recognizes such a setup would meet the monthly monitoring requirement as well as the automatic line leak detector requirement. The following conditions must be met:

- **Sump sensors used for piping interstitial monitoring must remain as close as practicable to the bottom of interstitial spaces being monitored.**
- **Monthly monitoring records must be maintained for at least one year.**
- **Electronic and mechanical components of the system, including shutoff devices, sensors, pressure or vacuum monitors, must be tested annually for proper operation. Records of the test must be maintained for three years.**
- **Containment sumps that are part of the piping interstitial monitoring system must be tested at least once every three years for liquid tightness. Keep the results for at least three years.**

Line Tightness Testing

During a line tightness test, the line is taken out of service and usually pressurized above the normal operating pressure. A drop in pressure over time, usually an hour or more, suggests a possible leak. Suction lines are not pressurized very much during a tightness test of 7 to 15 pounds per square inch.

Most line tightness tests are performed by a testing company. You should observe the test. Some tank tightness test methods can be performed to include a tightness test of the connected piping. For most line tightness tests, no permanent equipment is installed.

In the event of trapped vapor pockets, it may be impossible to conduct a valid line tightness test. There is no way to tell definitely before the test begins if this will be a problem, but long complicated piping runs with many risers and dead ends are more likely to have vapor pockets.

Some permanently installed electronic systems, which often include electronic line leak detectors connected to an ATG system, may meet the requirements of monthly monitoring or a line tightness test.

Check with your implementing agency to determine what is allowed.

Links For More Information

Government Links

- U.S. Environmental Protection Agency's Office of Underground Storage Tanks: www.epa.gov/ust. EPA's UST compliance assistance: www.epa.gov/ust/resources-ust-owners-and-operators
- State UST program contact information: www.epa.gov/ust/underground-storage-tank-ust-contacts#states
- Tanks Subcommittee of the Association of State and Territorial Solid Waste Management Officials (ASTSWMO): www.astswmo.org
- New England Interstate Water Pollution Control Commission (NEIWPCC): www.neiwpcc.org

Industry Codes And Standards

www.epa.gov/ust/underground-storage-tanks-usts-laws-regulations#code

Other Organizations To Contact For UST Information

http://nwglde.org/